Halbritter's Plant-and-Animal World

Also by Kurt Halbritter

Halbritter's Arms through the Ages

Halbritter's
Plant-and-Animal World

Being a Modest Contribution to Natural History
for People from all Walks of Life
With Lots of Illustrations by the Author

Translated by Joanne Turnbull
with the assistance of John Githens

Seaver Books   New York

Originally published by Carl Hanser Verlag, Munich, Germany,
under the title Halbritters Tier-und Pflanzenwelt
Copyright © 1975 Carl Hanser Verlag München Wien

First Edition 1981
First Printing 1981
ISBN: 0-394-51805-5
Seaver Books ISBN: 0-86579-011-6
Library of Congress Catalog Number: 81-2277

Library of Congress Cataloging in Publication Data
Halbritter, Kurt.
Halbritter's Plant-and-animal world.
Translation of: Halbritters Tier- und Pflanzenwelt.
Includes index.
1. Animals, Mythical—Caricatures and cartoons.
2. German wit and humor, Pictorial.   I. Title.
II. Title: Plant-and-animal world.
NC1509.H3A4 1981   741.5'943   81-2277
AACR2
ISBN 0-394-51805-5

Manufactured in the United States of America
Distributed by Grove Press, Inc., New York
Seaver Books, *333 Central Park West, New York, N.Y. 10025*

Part I • Animal World

Pointer
demonstrator simulans

Water bird which can neither fly nor swim, and so never goes in over its head. Principal habitats: the Handalusian Coast, the Bight of Thum. The Pointer is aptly named, since it seems to point sometimes in one direction, sometimes in another, and sometimes all over the place.

High-handed Chucklefoot
pes ridiculus cothurnatus

Fleet-footed land bird. Lives in swamps and marshes, where it buries its eggs away so ingeniously that it can never dig them up again. When mating, this bird bursts out laughing—a trait which distinguishes it from all others.

Water Thumber

pollex aquarius

A bird with fingers for wings that barely gets off the ground, but is
nonetheless agile on the water and under it. Though known for its
shyness, the Water Thumber is prey to an insatiable curiosity, often
causing it to fall into the hands of hunters.

Stopcock
claustrum cannale

Water bird of Chlorania. Lives on water alone, water rich in calcium and chlorine for the most part. Consumes vast amounts of the aforesaid and where the species abounds, so do water shortages.

Prayer (or Beggar-bird)
precator sollicitus

Known as a bird in the hand, this bird would rather be in the bush, since it is scared to death of the water on which it lives. The young are carried around by hand and only learn to live in the water when they are older. A capsized Beggar-bird is a dead duck.

Finguin
finguis index

Grounded water bird never bitter about its fate. The Finguin will flip its fingerbill at any bird who tries to step on its toes. Among its own kind, however, the Finguin is generally a perfect gentleman. A very social bird, the Finguin bows by way of greeting fellow Finguins. Finguins never baby their babies: they lay their eggs to the wind.

Long-billed Stovepiper
acrocephalus fornax

Can't resist rust or the draft(s). In ancient times, the Long-billed Stovepiper was worshiped like a god and given burnt offerings in the hopes that the winters might warm up. The bird never rose to the occasion and is slowly dying out.

Footard
futeo furans

Marauding nest-wrecker. Foot-bound yet footloose, crazy about crags, less crazy about the water but a pro in the air. This feckless footster builds its own nests but takes off with the eggs from others in tow: too lazy to hatch its own. Huge appetite.

Nipperwill
forceps mordax
A small, shy bird with real pluck. Savagely seizes upon anything
sticking out one inch or less. At home in Tonga and against the
Tongan labor laws, little Nipperwills are caught and made to work
as minors, throwing a wrench into the Tongan tool works. Nipperwills can live twelve years and pry loose eight to nine hundred nails
daily, plus tacks and brads of all descriptions.

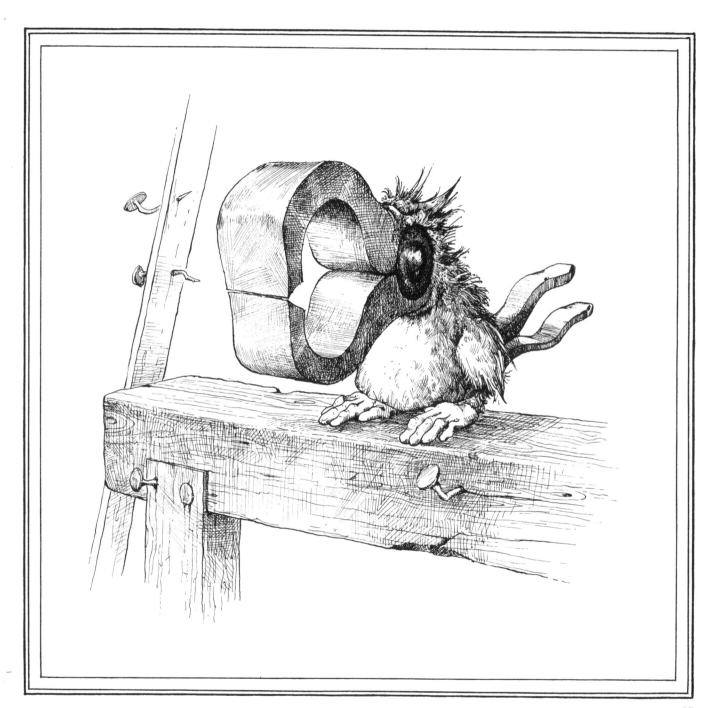

Drill-billed Owl
ulula terebrans

Alias the Dentist Owl, because of the furious pace at which it bores holes into anything that comes near its screwy beak. The beaks of twenty-five-year-old Drill-bills may attain diameters of more than half an inch. Properly trained, Drill-bills make terrifically useful pets for around the house and in the garden.

A-screw-loose Bird
rostra vertiginosa

Sprightly, energetic, "the Carpenter's Nightmare." This feather-brain feeds on what's left in the holes made by screws, which it deftly undoes with a twist of its bill. Its natural enemy is the left-handed screw.

Light-footed Pipe Puffin
fumatorium foetidum

Habitually goes up in smoke. Rare bird from Briarland; feeds on dry leaves and fumes a lot; supposed to be shy, but lightning-quick and very curious. The Pipe Puffin is the type, if frightened, to put panic in its pipe and smoke it: the best protection around. House-broken, this perennial puffer proves the ideal pet: a constant conversation piece if not always useful.

Snatchclaw

accipiter accipiter

The Snatchclaw never sleeps. Ever-greedy, it will grab whatever it can get its hands on. Lets go only what it cannot eat, and then only grudgingly. A taker by nature, the Snatchclaw lets the other birds do all the work building the nest, then steals the show. Originally from Hants, this bird has been brought to these parts and put to work as a procurer for the lowly elements in high society. Which explains why it is the prototype for the beast on the local coat-of-arms.

Chimney Swift
apus exstructus

High-flying hotshot given to sweeping down on castle ruins and chimney tops for a good time. Can wing it alone anywhere, no matter how steep the cliff, how old the castle, how high the wall. Loves to eat and run: plucks mortar and worms from crumbling castle walls while in flight. A Natural Wonder.

Minster Sparrow
passer cathedralis

Sadly dying out due to the recent decline in church construction. Unlike so many today, these birds still have Faith and wherever one finds an ancient cathedral, spires pointing heavenward, one is sure to find the Minster Sparrow as well. It roosts on top of the gargoyles, plugging up the spouts to protect the baby Minsters from becoming fallen Minsters. Not to be confused with the Romanesque *(passer romanus)* or the Baroque Sparrow *(passer perversus),* which are found mainly in the North.

Great-eared Gawk
lectulula auriculata

Flies by night, hears with difficulty, and is virtually blinded by the light, yet cheerful withal. Home is Earan. Has lived in trees ever since being banned from the bedroom by the interdict of 1792. Its cry is ear-piercing if not heart-rending; its behavior strange. No peasant wench is safe alone in bed at night and the Gawk's forays are rarely without consequence: hence its nickname "Eerie."

Lampyr

vespertilio voltarius

Highly charged energy wasters: consume anywhere between 110 and 220 volts daily. Prefer to plug in at night for best results. Lampyrs are the vilest among the vile volt-suckers, and hardly harmless. Contact with these nocturnal no-goods in the presence of high humidity can be fatal. If attacked, they will cause the circuits to short and make their getaway in the pitch dark.

Bulbous-browed Tortoise
testudo candens

Harmless lampshader, sluggish and extremely reticent. Tamed, the Tortoise casts a cozy glow over bed or night table. Known for the neon glow of their bright if not illuminating minds, they are easy to spot at a distance. Originally from Mazdastan.

Voltasaur
dinosaurus electricus

Power-line feeder, adapted for 110 volts. Due to the great demand in recent times for these energy-efficient miracle workers, they are now in perilously short supply, surviving only in small numbers at the southern tip of Plugsend. Labeled an endangered species, Voltasaurs are on the EPA's top priority list. Voltasaurs light up at the thought of making love, a real turn-on for both sexes: the males get red, the females green. Voltasaurs are highly charged with a strong libido which makes them prone to all those uncontrollable urges when excited. Not known for their subtle approach, the incandescent glow of their dorsal bulbs lets the whole world know what makes these reptiles tick.

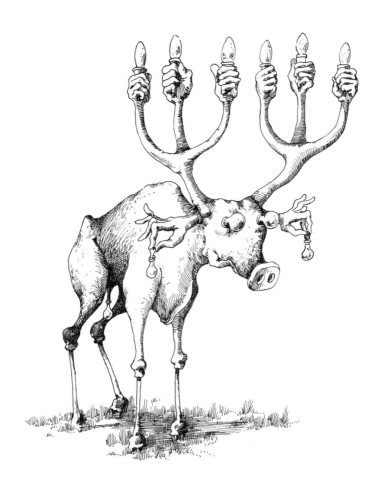

Polydactyl Blinkbok (or Chandeleer)
cervus senibrachialis fulgens
Feeds exclusively on currents. Tends to get lightheaded as the mating season comes on. Can be found roaming the western plains of Mazdastan, also in Lampardy. Flashes for right and left turns; switches off in cases of emergency.

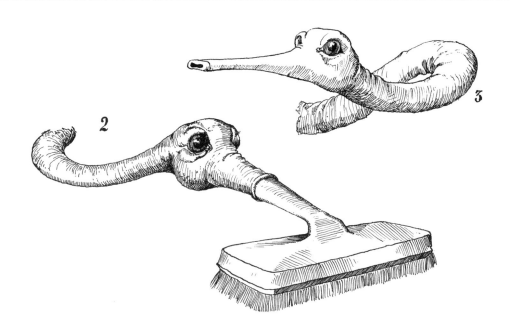

Hoover-has-it-all
purgator sugens vulgaris

All-purpose dust sucker from Kontacta, clings to any carpet and won't let go. Clean and safe, this domesticated Disposall is a great favorite among modern housewives. Once you have yours, you won't know how you ever lived without it. The new Hoover-has-it-all comes in three deluxe models:

Regular (1)
purgator sugens vulgaris
Does the usual job in a jiffy.

Bristling (2)
purgator sugens peniculatus
For those tough-to-clean stains.

Extra-long (3)
purgator sugens rostratus
For those hard-to-get-at corners.

Tyre Back
cingulum rotans
Very quick and agile. Consumes up to 500 miles a day. Tends to show signs of wear and tear with time. Tyre Backs, or Tareaters, travel in gangs and are street-fighters by nature.

Grubble
camelopardalis avida
Originally from Excavasia but, sweet-tempered and adaptable, Grubbles are happy and at home wherever there is something worth scrounging up. Fully grown Grubbles grub up from ten to twelve tons daily; trained at an early age, they make the best rubble removers around and are invaluable to the road construction crews in technically backward Excavasia.

Graffitie (Fabermouse)
plumbaculum faberi

Lives on paper; given a little guidance can do the most amazing things; distinct daytime and nighttime habits. Produces, when bored, stick-figures and doodles of all denominations. Original habitat: East and Central Grafica.

Fingeroo
halmaturus digitatus

A real handful, the Fingeroo cannot stop jumping around. Capable of enormous leaps and bounds in any and every direction. Although volatile by nature, the Fingeroo is a faithful lover and picks his partner according to the size of her pouch.

Horsehound
yaenula pedester

Originally from the Solomites. Delicate and fleet-footed; hunts
only at night. Exceedingly timid and shy. Lives alone in regions
that are completely uninhabited and inaccessible. Emerges briefly
for the mating season and at the risk of being hounded.

43

Deerhart (or Tenderleer)
decicornua periurans

Shy and sweet but stubborn. The Deerhart has a strange habit of
putting itself under oath in case of danger, perhaps because it truly
believes that justice will win out in the end. Native of the northern
reaches of Footsieyama and North Handorado, on Cape Handy.

Handshovel Elk
alces exactor

Roams the plains in herds. Firmly convinced, unlike the Deerhart,
that might makes right. This Elk's handshovel always hangs open
and pounces, as a rule, on every twentieth animal that has the
misfortune to cross its path.

Leghornucopia
illex iambecornuta

Indigenous to the Solomites. Long-haired, sure-footed and provocatively crowned with a pair of ravishing legs. The luscious Leghornucopia lives way up in the mountains, at an altitude ranging anywhere from 8,000 to 12,000 feet, and comes down from these peaks only on the coldest winter days. The severe climate of its natural habitat notwithstanding, this erotic beast has led many a hunter astray, enticing and exciting him to such heights that they can hardly hold their peashooters straight.

47

Foothorn Goat
capretta pedicornuta

A finger or toe in every pie, this goat has a handle on everything.
Staggeringly smelly and naturally combative. Will throw its feet up
in the face of any animal that tries to challenge it: the best defense.
Female Foothorns are endowed with an udderhand between their
legs with which they coyly salute the male population. Originally
from Footsieyama.

Barefooter
pedarius nudus peregrinatorum
Found in the Solesbury Marshes when not out on the road. Sure-footed, smooth-running and subject to serious bouts of wanderlust. Has a good heart and loves people: a natural for pilgrimages and religious processions of all kind. Frequently seen taxiing home from Mecca or Lourdes, having given a lift to the devout souls it met there. Diet consists mainly of leaves, moss and leftovers from the pilgrims' picnic suppers.

Two-hander
operarius duplex
One-hander
operarius simplex

Although the Two-hander and the One-hander are related to each other, they live in different places: the latter in Footsieyama and Inner Monsolea, the former in Outer Monsolea. In spite of the distances, they share a strong family feeling and do a lot more than just shake hands when they run into each other at the territorial borders. These kissing cousins are fun-loving and not a little promiscuous. Easily domesticated, they make excellent handymen.

Greater Earcocker
auscultator magnus

Not to be confused with the Lesser or Dwarf Earcocker *(auscultator minutus),* the Greater strain grows to be much taller (10 to 12
feet). Thanks to the fantastic size of its ears, this mammoth creature can hear a pin drop. During the mating season, huge herds of
Greaters go out to the Earals of Outer Monsolea looking for action.
The boys (bulls) get together to tussle and trunk-wrestle over the
girls (cows) who stay on the sidelines, watching closely and keeping
score.

Breastling

quadrupes mamillata

A fat, four-footed, meat-eating mammal, short and squat with overstuffed legs, chubby feet, bulging eyes, furrowed brows and pendulous ears. To protect itself, the Breastling lunges breastlong into the enemy, its powerful ears erect. Physically well-developed, a keen sense of smell and touch, but mentally not abreast of it. The King of the Busts.

57

Double-horned Wishbone
osseus cornutus

Bone-poor native of Rheumatia. Scuttles about the boneyards of Verterbra in search of real marrow, its staple food and the only thing that`makes life worth living. The female lays two bones a year.

Rackaribs
turbo intervertebralis

Found exclusively in the badlands of Rheumatia. In the dry season, does nothing but trudge sadly across the dunes, dreaming of less dreary days. A forlorn fluke of nature showing no signs of life, it remains a mystery to us all how the Rackaribs ekes out its existence. A wide-open field for biologists.

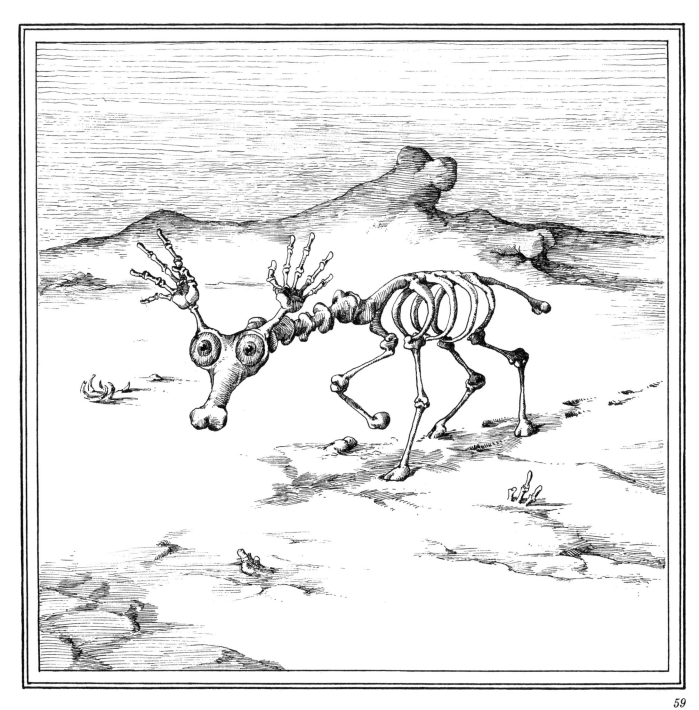

Nosehare
lepus nasutus

Distant relation of the Fingeroo (see page 41), but smaller and less outgoing. Can generally smell a rat before it sees it and a hunter's horn before it hears it. Harebrained animal that one is best advised to shoot only on sight. Tamed, it is no substitute for dogs or cats.

Snifflemouse
nus olfactans

Distant cousin of both Church and Lay mouse (kitchen variety). Unlike the latter, it doesn't mouse around but sticks its nose into everything. Its abnormally enlarged olfactory organ (the sniffle) makes it the nosiest creature on earth. It sniffles loudly, especially at night, particularly annoying to anyone trying to get a good night's rest. Snifflemice, however, never cry.

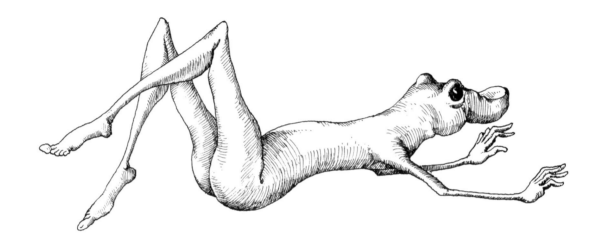

Hip-hopper
gibber resaliens

A grasshopper with great legs: sexy and saltatorial. Can leap as high as 14 feet, 3 inches—no mean feat for an insect just 4 inches tall and weighing less than an ounce.

Giant Handrover
palma dodrantalis

Nearly extinct thanks to natural selection. Only a few surviving in South Handalusia, near Handong. Hopelessly myopic, this meandering misfit is all thumbs and can never get a grip on anything except by accident. Also known as the Big Bumbler. Fortunately for us, the dwarf variety has never existed.

Pokenose
pollex sollicitans

Pokes its nose in everywhere you wish it wouldn't, pleasant but persistent, sweet but insatiable. Has discriminating tastes but looks down its nose at no one, so is loved by all. A much better friend than most. In Thumatia, no lady of refinement is without one.

65

Feelfree

digitulus fraudolentus

Indigenous to Central Handalusia. Always feels fine and lives seventy years on the average. Procreates by sleight of hand. Indulges in little Boy Scouts for lunch.

Hand-to-mouth

manus timida

The Hand-to-mouth is on its last legs with only a few still left in Handalusia. Exceedingly introverted and painfully shy; lives alone since lovers are hard to find. Zoologists agree that at this rate the Hand-to-mouth can't last much longer.

Droop-tailed Ballscratcher
penicillum flaccum

No ball of fire but nevertheless plays hard to get. Has been spotted only rarely and then only in the Palps Valley. Despite years of research, scientists are still baffled about where little Ballscratchers come from. Not very big on the uptake, mature Ballscratchers are known for their poor posture, but they will straighten up in a minute if excited or challenged by something: a dramatic habit which frightens the daylights out of their would-be detractors.

Fat-tailed Faker
praeputium impeditum

Lumbering mammal of limited intelligence. Physically unprepossessing, stockily built and pinheaded. If frightened, will hide its head away in the voluminous dewlaps that hang from its neck and hope for the best. Its full extension is supposed to be 13 inches but it rarely, if ever, gets that far. Could hardly be called quick on the draw which is why it never reacts to lures or decoys, let alone the real thing. Lives a sheltered life at the foot of the Handes.

River-tit
papilla sypho

Lives at the mouth of the Bazoomi, near St. Bibulus. Squirts at flies passing overhead, both the zip-up and button-down kinds, the key staples in the River-tit's diet. River-tits figure heavily in Bibulusian landscape architecture: used as fountains that are beautiful to look at and functional too, guaranteed to keep any garden fly-free till the day they die. Only Bibulusians know how to kill two flies with one tit.

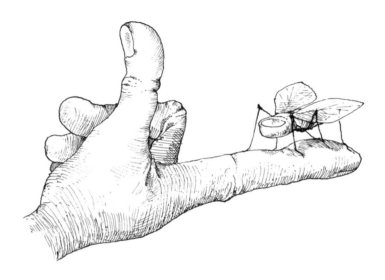

Tie Fly
musca focalita
Fingertip Fly
musca unguiculata
Both flies fly and make a maddening racket in the process which is
continual; can drive any man or beast insane and do so constantly.
Indeed, these persistent pests have caused so many to fly off the
handle as to have no friends left: they fly, fly, and fly again but, no
matter, they always fly alone and die a lonely death, eaten, sprayed
or swatted.

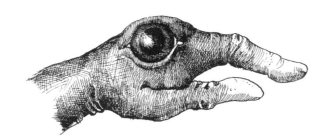

Flatfoot
planta odorata
Puts its feet up at home in the Solesbury Marshes. Is smelly and
not very arch; antisocial and so unpopular as to have been discon-
tinued in most places.

Polypedalpotamus
nasopes benignus

Slovenly, slaphappy mammal stuck in the mud of the Solesbury Marshes more often than not. Roughly 11 feet long and 5 feet tall, pleasantly plump with a steadily tumbling tummy and an unfortunate nose. Lives with its foot forever in its mouth. Peaceloving and sweet-tempered but otherwise a stinker. Never ready to begin the day, it has to kick itself with its kicktail to get out of bed in the morning. Dried Polypedalpotamus dung is used as insulation in low-income housing everywhere.

Fingerasp
natrix unguiculata

The Fingerasp doesn't bite but spits with a vengeance. Very badly brought up, probably from a broken home, and best avoided by those who behave themselves.

Linteaters
peniculum saetigerum
Ultra-democratic dusters, famous for their fastidious feasting and waste-not-want-not approach to every speck of lint that appears on their plates or near their palates. Found wherever better business suits are sold. The housewife's dream-come-true in two deluxe models: the basic bristle-back *(peniculum saetigerum dorsale)* and the terrific tummy-tickler *(peniculum saetigerum abdominale)*. Exceedingly affectionate; either one will win your heart, if not your lint.

Bootsnail
cochlea caligulata
Renew its sole and the Bootsnail goes on faith: 24 feet an hour.
Found in Inner and Outer Monsolea. Comes in two colors: tan or
black.

Creeppeeper
oculus cochleatus
Nightcrawler, 3 inches long. Has a very closed personality, won't
come out of its shell except at night. So unappreciative that once it
has become accustomed to the darkness, it can do without it. An
evil eye par excellence, it is big on withering glances.

Water Mouse
musculus aquatilis
Loves a good game of cat and mouse. A popular player because
such a good loser: the Catfish can't complain.

Catfish
felis inutilis
Pretty fishy cat-of-all-trades, found sometimes in the depths and
sometimes on top of the Puss Sea. Pushy like its cousin the House
cat, it rules the roost when it comes to mice. According to Professor
Catcalls, Catfish tend to exaggerate insofar as they are "a cross
between too many cats and more cats than necessary." But then,
that is a cat of a different color.

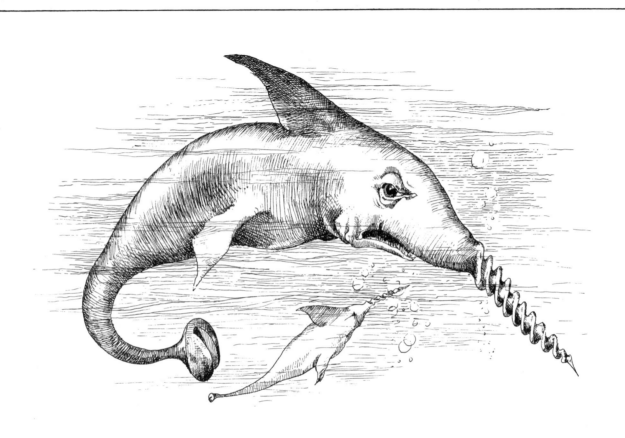

Turn-of-the-screw Shark
pristis cochleata vulgaris

Has a long snout ending in a contaminated corkscrew, considered
deadly. Not a wine lover and naturally combative: hangs around in
schools after classes are over. Becomes vicious for no apparent reason and attacks by furiously screwing up its nose at its victim.
Always mean and nasty, it will screw anything on fins or off—fish-
ing boats included. Has a way of dragging them all down to its own
low level.

Cunning Cuddlefish

octobrachium cupidissimum

Sleazy eight-armed sea monster; can't keep its hands off anybody.
Polydactyl and puritanical, it preys primarily on Christian sailors
who have gone astray. Research into the sex life of the Cunning
Cuddlefish is currently at a standstill for lack of any captive speci-
mens. Cuddlefish cuddling is a tough act to catch, let alone follow.

Sea Sprinkler (or Little Squirt)
hippocampus bisitulatus
Sweet, endearing sea creature indigenous to the coastal waters off
the Soakes. Adorable, docile and dutiful, the Sea Sprinkler is a real
asset in the right hands: it squirts every time. As the saying goes, "a
sprinkler in the hand is worth two squirts in the bush."

Boggle-eyed Bumwiggler

saltator clunium perspicillatus

Liberal-minded with a palatial posterior: will wiggle its round rump at whatever comes along. A real tease, it typically gets in over its head. Usually found at the bottom of the Callipygian, drinking in the perfume of the Fingerrose, essential to its diet. The Bumwiggler's worst enemy is the Turn-of-the-screw Shark (see page 84), who can't stand a come-on.

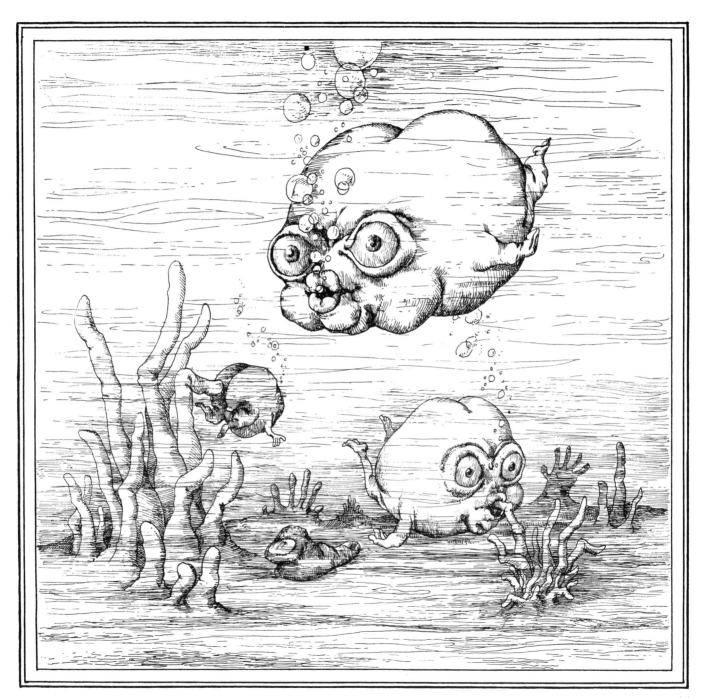

Sea Manikins
masculus marinus

Otherwise known as Automated Malemaids, their tails are fitted with propellers that propel them through the water in a tenth the time it takes by hand. Members of the modern generation, Sea Manikins can and do keep abreast of any Jelly-tit worth its salt.

Jelly-tit
medusa mammata

Free-floating and burgeoning with budding balloons that bob up and down. Eats with its hair in its face and chews with its mouth open. Jelly-tits cannot keep anything to themselves and, if at all excited, will give anyone the juice—especially Sea Manikins and Boggle-eyed Bumwigglers (see page 90).

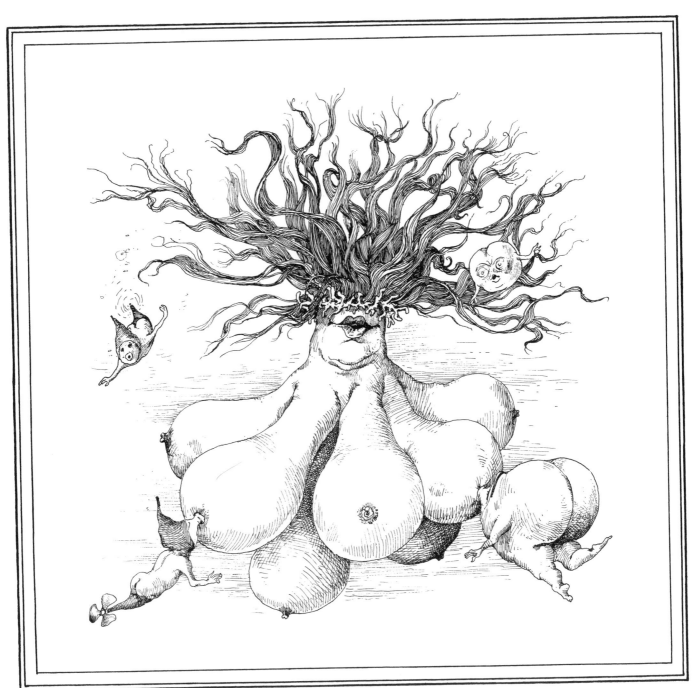

Part II • Plant World

Peeproom (or Shuteye)
morchella oculata

Hollow-legged fungus with a well developed visual sense. Looks around all day, every day, and shuts up shop at night. Very perceptive with a fine sense of humor: always sees the lighter side. Abounds in woods and visual fields, especially those by Seabury Wells. Has a good eye even on a bad day.

Feather Caps
paxillus rupicapratus
Also known as Yodelids, these headhunters hang their hats in the
Hatrack Valley. Edible but tough and bitter even after hours of
chewing. Feather Caps are notorious for bad taste.

Frothfrought
boletus spumatus
A fungus always foaming at the mouth and, despite its unsightly
ways, a favorite round the world. It has yet to be established where
the Frothfrought first sprang up, but recent studies suggest the
Bliss-on-Barleycorn area.

Double-dealing Devilstool
sarcodon diabolicus caudatus

Mean and mischievous, this fungus has a devil-may-care attitude and horns in everywhere. Rare, but aptly named. One should always judge this fungus by its cover: the horns say it all. Skillfully concealed at the foot of a tree, the Devilstool reaches out and trips up the unsuspecting victim with its hellishly long appendage. Absolutely apolitical, it reacts to right- and left-wing radicals in the same diabolical manner and inspires the fear of hell in a majority of theologians.

101

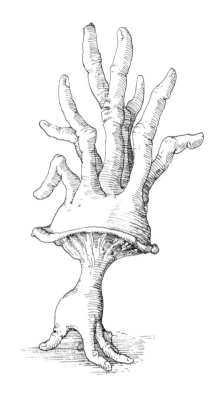

Finger-twiddling Fiddlecaps
boletus iocosus
This quick-sprouting cousin of the Titillating Toadstool has the fair
sex all excited. Found in the bottomlands of the Missusslippi, it
twiddles its fingers and fiddles under the skirts of many a 'Slippian
Miss, those meek young mushroom-gathering maidens who go looking for it. Men have no idea.

Plushstool
paxillus holosericus

Eighteen inches high, 24 in diameter. Provides the natural way to relax and unwind after a tense day in the fields. Plush, plump and palatial, this heavenly hassock will accommodate even the amplest of posteriors. Indigenous to the Breastwolds, the Plushstool is a real hit with the larger local peasantry. They have even named their "Plushstool Polka" after this fabulous fungus.

Foot-fetish Fungus
boletus noli me titillare
Crops up in the foothills of Solensk. A delicate comestible, the handling of which poses an exceedingly sensitive problem for importers. The slightest touch can throw this fungus into convulsions, ruining its subtle flavor forever.

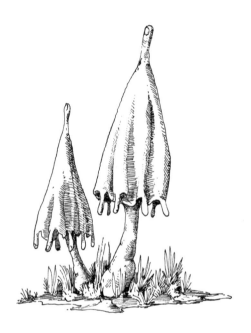

Rainy-day Friend
refugium boletum
Indigenous to Umbrella Land. This foul-weather fungus is meant
to be seen and not eaten as it will provide shelter for any traveler
caught in the rain. Far better than any synthetic slicker you can
buy, this free-for-all, natural fungus comes with collapsible cap and
built-in windbreaker.

Hand-carved Stonestools
boletus inedulis

The original fungus, various forms of which have sprung up all over the globe. A noncomestible considered sacred by the ancient Greeks, Romans and Egyptians, it has consistently resisted attempts to create an edible kind. The enormous difference in quality between the real and imitation Stonestools is indisputable. The plastic and Pyrex varieties are of no value whatsoever to the serious collector of antique Stonestools—the same goes for the rubber and Corningware versions. The genuine existing Stonestools are the following:

German Stonestool (1)
boletus inedulis germanicus

Egyptian Stonestool (2)
boletus inedulis aegyptius

Archaic Stonestool (3)
boletus inedulis archaicus

Gothic Stonestool (4)
boletus inedulis goticus

Baroque Stonestool (5)
boletus inedulis perversus

Kidney Stonestool (6)
boletus inedulis calculus

Bauhaus Stonestool (7)
boletus inedulis academicus

Saddle Stonestool (8)
boletus inedulis stragulatus

4

5

6

7

8

Botticelli Belly
boletus umbelicatus

Native to Navels, North and South, occasionally found in the voluptuous valleys of Ladydown and the sumptuous satin folds of Bedding. The smooth, plump, ripe flesh is delicious, juicy and sweet. An acquired taste, it is sometimes too rich for those who have never tried it before. Considered a great delicacy in Navels, it comes in three mouth-watering varieties:

Lady-button or **Female Belly (1)**
boletus umbelicatus femineus
Billy-button or **Male Belly (2)**
boletus umbelicatus masculus
Bellybutton au naturel (3)
boletus umbelicatus nudus

Fungee Over Easy
ovum speculativum
Found sunny-side up in the lightly salted hotbeds of the Yolkatan.
Hatches from Fairy-chicken eggs and is done in minutes under the
hot noonday sun or over medium on the Gas Range.

Giant Hooknosestool (or Schnazola)
boletus nasutus communis

Hails from the Reek Islands and is not to be confused with its
northern neighbor the Stinkhorn *(boletus nasutus septentrionalis)*
whose shape is more angular and smell more pungent and whose
natural habitat is in the Ohzholm. Can be fatal if swallowed during
the hay-fever season, but otherwise, and when fully ripe, is succulent and savorful: the perfect little something for those special
occasions. The Reeks, known for their generous and welcoming
ways, serve today, as they did in ancient times, Giant Hooknose-
stool au Gratin to all their guests.

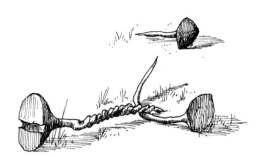

Screwlets (1)
nematoloma cochleatum
Tacky Fungee (2)
nematoloma clavissimum
Inedible products of industrial selection, these factory fungee have
survived because they are fittest, but maltreatment may still man-
gle their functional shapes. Native to the state of Rust, they are
not as resilient as their natural counterparts.

Flaky Flytrap
amanita muscata

From the wilds of Abuzzi. Cunning noncomestible, this catcher of
flies has so far eluded the human touch. The few who have suc-
ceeded in separating the flies from this fly-by-night fungus have
found no fungus at all.

Sweet Maidenmorel (or **Ladypop**)
morchella mamillata
A plump, luscious mushroom always popping out in new places.
Divinely sweet and sensual, it is a potent aphrodisiac for those who
indulge. Makes a delectable first course or, topped generously with
whipped cream, an irresistible dessert.

Finger Flower
margerita lacrimata

Grows to a height of 9 feet. An inquisitive cousin of the sunflower whose single eye stares into both of yours from the midst of a dazzling array of identical digits. Taken as directed, Finger-flower seeds cause the pupils to dilate, lending an undeniable twinkle to the eyes, no matter how dull the subject.

Lip Service
paxillus mordax

Offensive, fast-talking bigmouth, indigenous to the Canines of Prosthesia. Dangerous if not deadly, it will snap at anything that comes within spitting distance. This mad masticator has laid lip over everything from nuns to nonbelievers. Reaches its predatory peak between August 19th and September 22nd. Claims hundreds of lives every year.

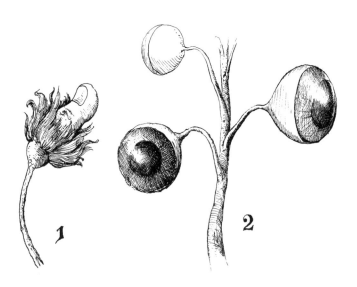

Thumbs-up (1)
herba pollica vulgaris

Multistemmed descendant of the dandelion with delicate, pink thumb blossoms pointing toward the sun in an overly optimistic fashion. Although Thumbs-ups are so sour smelling as to be off-putting even to the Foothorn Goat (see page 48), their petals make an excellent tea. Bitter Thumbs-up tea picks you up if you're down and keeps your warts from getting out of hand—perhaps the root of your depression.

Bug-eyed Blossom (or Peek-a-boo) (2)
erica pupillata tacita

Where this pretty little flower got its peepers nobody knows. Clear-eyed and curious, it peers across the Glimpsmoor and whatever it sees it keeps to itself. Extremely rare, this flower blooms for a day then closes its slightly bulging eyes forever.

Hoofer
folium ungulatum

This flower is often seen dancing in the wind, doing a little soft-shoe
with its hoof-blossoms. Lyrically light on its feet, this formidable
Hoofer is especially sought after for the design of its shoes.

Briarbrad
clavus pratensis

A coarse, prickly grass with a tough-as-nails stalk so strong it won't
bend even when it comes to blows. Carpenters collect these natural
wonders once they are in full bloom and easiest to pluck. Needless
to say, they are infinitely superior to the manmade imitations and,
as such, a perennially popular export item.

Playful Pucker-up
osculum fructuarium

Named for the shape of its petals, if not certain pronounced pro-
clivities. Sweet but not terribly subtle, this cousin of the onion can
hardly be called a late bloomer. Affectionate at birth, it begins
kissing before it can talk. This fantastic behavior was conclusively
confirmed last year by the distinguished biologist Dr. Con Amor,
whose daughter Anita happened to catch two Pucker-ups pucker-
ing up in the garden patch. Nine months later she had a healthy
little Pucker-up of her own. The bees claim no involvement.

Self-effacing Fingerweed
myriodigitus inutilis
A diversified plant composed of countless fingerlike formations
which get all in knots over one another, cut one another off, and
then drop off. The Fingerweed is the favorite snack of the Snatchclaw (see page 22), well known for its tendency to grab anything
not nailed down. Bewildering to the backpacker who usually beats
a hasty retreat while nervously glancing over his shoulder.

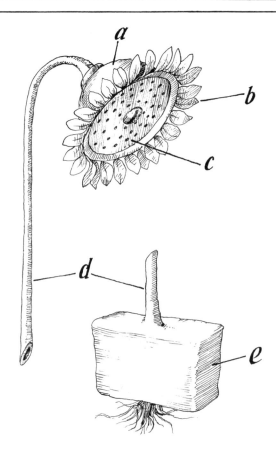

Sunshower
helianthus aestuans

First spotted on the banks of the Showabaden. The head (a) of the outdoor Sunshower is fringed with petals (b) and protected with a perforated leaf (c). The stalk (d) comes in all sizes up to 8 feet, and sprouts from a tap root (e) in which the water is stored. A shake of the stalk and you have a stimulating spring shower instantly, sweet and refreshing and not always just to the person taking it.

Wigweed
herba capillamentalis
A beautiful, bluish-green, threadlike lichen. Spun according to Dr. Purlman's special method, and given a grandmother who likes to knit, all-natural Wigweed can be made into snug, warm, winter hats to last you a lifetime.

139

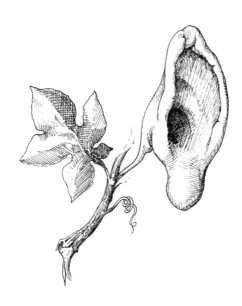

Earleander
vinca auriculata

A real earful, this exotic-looking fruit grows on the sunny slopes of Earleans in Earan. Freshly picked from the vine in the first fall months, its flesh is firm and tuggy and not overly juicy. What juice there is ferments and turns into a potent waxlike liqueur which everyone has heard about thanks to the ear-deafening ditties of Earanians under the influence at harvest time.

Udderly Amazing
frutex lactans

Leafy green tree bearing baglike fruit from which flows, under pressure, a magnificent milky nectar, nothing like the coconut's. Supplies are limited to the extent that each tree produces only a dozen amazings annually. Nevertheless, these Udderly Amazing trees have it all over cows since they never kick back.

Sugar Gams
crus dulcissimum

A teasing tropical grass that swishes in the wind. Slim and shapely, it tastes as good as it looks: the syrup inside is made for French Toast. Sugar Gams grow only in the foothills of Footsieyama. They are harvested with trip-wires and processed immediately to keep the flavor from escaping. Sugar-gam growers have been able to increase the quantities of syrup produced by as much as 25% by simply tickling the soles of the Sugar Gams' feet, three to four weeks before harvest. This process is known in certain agricultural circles as artificial intickleation.

Insidious Index (or Catchall Cactus)
cereus insidiosus erectus
Indigenous to Handalusia and prickly by nature. Admirers are best advised to keep their distance or at least keep their hands off. The Index looks harmless enough to the unsuspecting exercise freak or lone desert rat ambling across the dunes, but it will pounce instantly on either. Not a nice way to go.

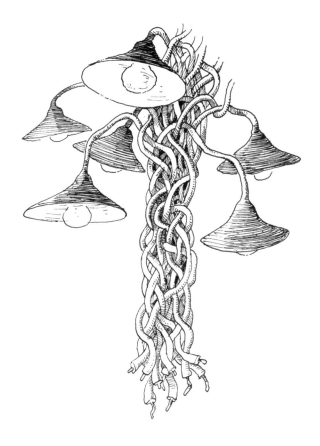

Wires-crossed Tangletree
arbor lucifera filiata
Native to Transformatia. Like its urban counterpart, the City
Lights (see page 156), it bears exceptionally bright fruit which cannot help but shed a little light on the situation. The only time that
the Tangletree fruit cannot enlighten someone is during the rainy
season when constant short-circuiting makes real communication
impossible.

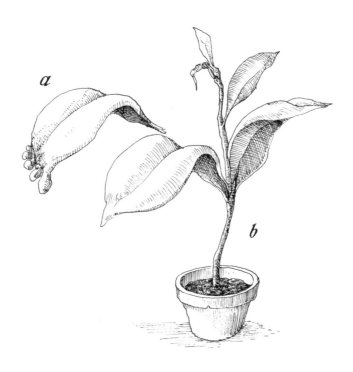

Flatfooted Palm
palma plantata
Wild and willful ancestor of the less ambitious Hothouse palm. The latter, not as colorful or courageous a character, prefers window ledges to wilderness, the known to the unknown. Indeed, the toes on this tame variety have all but disappeared since it never leaves home.

Finger Poplars
populus digitata

Reach a height of 18 feet and not much else. Though potentially gifted with their hands, these Poplars are shamelessly lazy and live exclusively on the handouts of petty Handalusian princelings who, for some petty reason that no one can grasp, love nothing better than to plant and tend these transparent parasites with their own hands.

153

Bosomberry Bush
pyrus mamillata

Smooth-barked and very well endowed, this tree grows to a height of 12 feet. Its pink-budded, flesh-colored fruit is sinfully succulent and functional too: as a child's pacifier or Christmas tree ornament. The ring-around-the-rosy Bosomberry Bush has completely eclipsed the maypole dances in these parts, despite minor casualties. The man who happens to be hit by a falling bosomberry is never quite the same again.

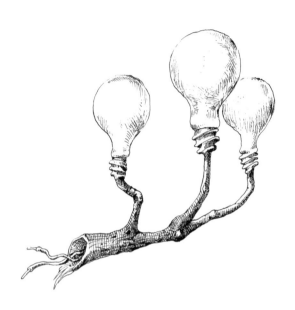

City Lights
candelabrum piratum

Unlike its country cousin, the Wires-crossed Tangletree (see page 148), the City Lights is terribly social, witty and urbane. Giving and gregarious, this extraordinary tree lights up at the thought of a little night life and therefore tends to attract it. Cafés, coffee shops, and restaurants crop up wherever they do. They run on 110 volts but can easily be adapted to 220 if exported outside of Mazdastan and Dynamonesia, their native habitat.

Knob-kneed Oak
quercus servilis

All knuckles and joints, this gnarled, rough-barked, hardwood tree
dates back to feudal times when men went everywhere in heavy
suits of armor. Needless to say, few of these obsolete oaks still
survive today now that everyone is dressing down.

Index

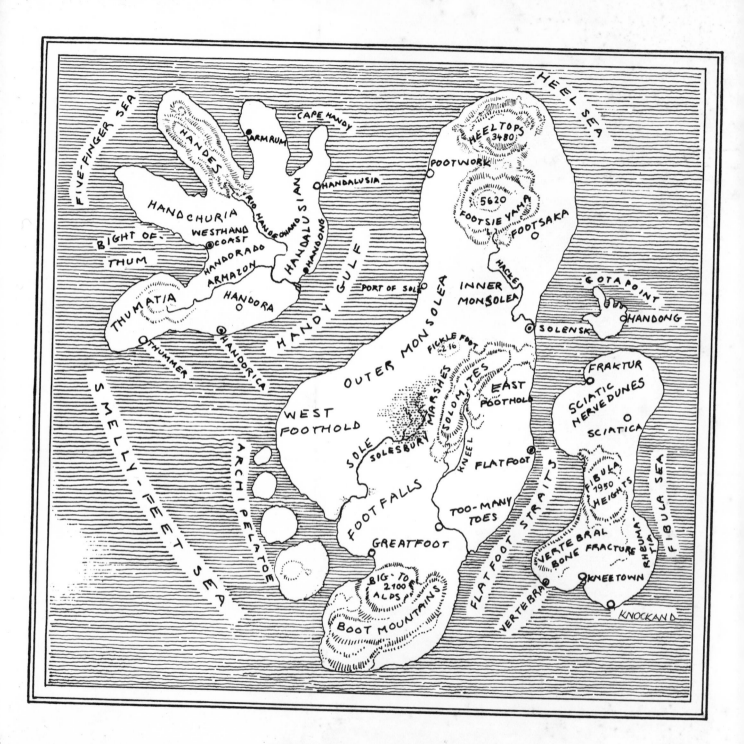